ideal
school supply

Grades K-2
ID43020

Funtastic FROGS™

AF700648

Word Problems

By Patricia Cartland Noble

Contents

Teacher's Note 3
Sample Solutions 7
Photocopy Master for Frogs 8
Activities 1–4: Look for a Pattern 9
Activities 5–9: Use or Make a Picture 13
Activities 10–14: Make a Table 18
Activities 15–18: Make an Organized List 23
Activities 19–22: Use Logical Thinking 27
Photocopy Master for Lily Pad Number Line 31
Photocopy Master for Write Your Own Problem 32

Published by Ideal School Supply
an imprint of
McGraw-Hill
Children's Publishing

McGraw-Hill
Children's Publishing

A Division of The ***McGraw-Hill*** *Companies*

Published by Ideal School Supply
An imprint of McGraw-Hill Children's Publishing

Editor: Jill Osofsky
Cover Concept and Page Design: Sara Mordecai
Cover Illustration: Kristin Mallory
Text Illustrations: Jim Connolly

Send all inquiries to:
McGraw-Hill Children's Publishing
3195 Wilson Drive NW
Grand Rapids, MI 49544

Funtastic Frogs™: Word Problems, Grades K–2
ISBN: 1-56451-360-2

2 3 4 5 6 7 8 9 05 04 03 02 01

Teacher's Notes

This book is one in a series of books designed to develop children's mathematical thinking. Each book supports the use of Funtastic Frog™ Counters in activities that teach key mathematical concepts.

Funtastic Frog™ Counters are available in six different colors and in three different sizes (3g, 6g, and 12g). The Funtastic Frog™ Counters are available from Ideal School Supply or your local Ideal School Supply dealer. Please visit our Web site at *www.MHteachers.com.*

In *Funtastic Frogs™: Word Problems, Grades K-2,* children use the frogs to:

- act out problems
- recognize and continue patterns
- use patterns to make predictions
- use a picture to clarify thinking
- gather and record information using a table
- make an organized list to organize combinations
- use logical reasoning
- model addition and subtraction

The activities in this book are designed to help children develop skills using different problem-solving strategies. The problems in these activities support the development of number sense and an understanding of operations. All the activities in this book support current mathematics standards.

Contents

This book contains 22 engaging activities divided into 5 sections. Each section contains similar types of problems, giving children the opportunity to practice and use each strategy.

Solutions are provided on page 7. A photocopy master for the frogs is included on page 8. A photocopy master for a Lily Pad Number Line is provided on page 31. Students can cut and tape the sections together and use it to solve problems. The Write Your Own Problem photocopy master on page 32 is provided for children to create their own word problems.

Math Skills and Understandings	Related Activities
Act out problems using objects	1–11, 13, 15, 16, 18–22
Recognize and continue a pattern	1–4, 7, 8
Make predictions	2–4, 8
Use a picture	1, 5–9, 12, 19, 20
Make a table	2, 4, 10–14
Make an organized list	15–18
Use logical thinking	5, 19–21
Write number sentences	3, 6, 9, 11, 12, 18, 21, 22
Solve addition and subtraction problems	3, 5, 6, 9–12, 14, 18, 21, 22

Suggestions for Classroom Use

The activities are sequenced by level of difficulty within each section and from section to section. Modify a section if you find it is too challenging for your children, or not challenging enough. The activities may be introduced to the whole class using an overhead projector, the chalkboard, or sitting in a circle. Once children understand the directions for the activities, they can work in pairs, small groups, or individually at a learning center.

Encourage children to share their thinking with the whole class. Children in the same classroom may possess different levels of mathematical understanding. What is difficult for one child may be simple for another. Talking about how to solve a problem allows one child to clarify their thinking, while for others it presents an opportunity to learn new skills to solve similar problems. Discussions about problem solving help children make mathematical connections and use the language of mathematics.

After students have completed a section, it is especially valuable to have them create their own problems. Students love to be creative and challenge one another in this way. Often times, it takes more thought to write a good problem than it may to solve one! Working in pairs or individually, students can create a class set of problems using the photocopy master on page 32.

Materials

For each child, pair of children, or group, you will need:

- pencils
- crayons or colored markers to match the colors of the frogs
- a tub of frogs in three sizes and six colors
- an activity sheet

Introducing the Activities

Before beginning the activities in this book, it is important to allow time for children to freely explore the frog counters. This will allow them to satisfy their curiosity about the frogs before using them in the activities.

When you introduce a section of the book, lead the children through the first activity in that section. Discuss the directions and how to record their work. You may wish to discuss the strategy or strategies they will use to solve the problem. To help students work through a problem, suggest that they use a four-step approach:

1. **Understand the Problem.** Have students retell the problem in their own words, clarify vocabulary, and state what is being asked.
2. **Make a Plan.** Ask, *What strategy is helpful?*
3. **Carry Out the Plan**
4. **Check.** Ask, *Have I answered the question? Does the answer make sense?*

You will find that although the sections are organized by strategy, an activity will often incorporate the use of two, or even three strategies to solve a problem. For example, students may use the strategy *look for a pattern* as well as *use a picture* to help them predict what comes next. When students use a picture to solve a problem, they also *use logical thinking* to figure out what the picture should look like.

Questions involving basic addition and subtraction are included on most activity pages. Students are asked to write number sentences to express the number relationships in the problems.

When reviewing the activity with the group, encourage the children to talk about their discoveries. This gives students opportunities to hear new ideas and compare them to their own thinking. This also provides an opportunity to determine which strategies are useful in different situations. As children explain their solutions they can often discover computational errors and self-correct. Furthermore, discussion allows you to better assess the level of children's thinking and understanding.

Mathematical Content by Section

Activities 1–4: Look for a pattern. While many children are familiar with repeating patterns (red, blue, red, blue), the activities in this section introduce children to growing patterns.

Activities 5–10: Use a picture. Students use reasoning to help them determine where the frogs' "mailboxes" are located. They also use the Lily Pad Number Line (page 31) to solve problems involving comparing numbers and skip counting. Be sure students understand how to count hops on the number line.

Activities 11–15: Make a table. Students begin this section by making an array of frogs that organizes them by color and size. Next, they use tallies to keep track of objects in a picture. Students are also exposed to using a table to make combinations. Finally, they complete a table that is fairly complex, but can be easily modeled or acted out.

Activities 16–19: Make an organized list. Students design toadstool caps, help frogs share a lily pad, and decide how many bugs a frog catches by finding all the different ways to combine two objects from a set. Next, they move on to discover arrangements of three objects.

Activities 21–23: Use logical thinking. Students use reasoning to help them determine the number of frogs in various settings. They also use clues to identify members of frog sets. Review the terms *more than, less than,* and *same number of.* Children then write number sentences to add and compare quantities.

Sample Solutions

These are sample solutions. Additional solutions are possible for some activities.

Activity

1 **A** 8 **B** 20

2

Hop	1st	2nd	3rd	4th	5th	6th
Rock	2	4	6	8	10	12

DO MORE 7 hops

3 **A** 5 **B** 1+2+3+4+5=15

4

Row	Number of frogs
1	1
2	2
3	4
4	6
5	8

DO MORE 21

5

Blue	Yellow	Orange
Green	Purple	Red

DO MORE Green Frog; Orange Frog; 9; 4

6 **A** 18 + 2 = 20 **B** 12 − 2 = 10 **C** 7 − 4 = 3

DO MORE Answers will vary.

7 **A** 3, 6, 9, 12, 15, 18, 21, 24 **B** 4, 8, 12, 16, 20, 24

DO MORE 5, 10, 15, 20; Red Frog, 15; Blue Frog, 20

8 12, 18, 24

9 **A** 12 + 8 = 2 **B** 10 − 3 = 7

C 12 − 10 = 2 **D** 7 + 11 = 18

10 **A** 9 **B** 3 **C** 3 **D** 5

11 **A** 7 + 2 = 9 **B** 3 + 6 = 9

C 7 − 2 = 5 **D** 7 + 3 + 6 + 2 = 18

12 **A** 7 **B** 13 + 10 = 23 **C** 13 − 7 = 6

13

	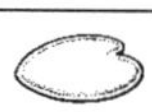	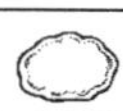	
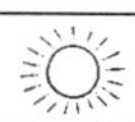	X	X	X
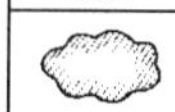	X	X	X

He can hop 6 places.

Activity

14

	1 minute	2 minutes	3 minutes	4 minutes	5 minutes
Small	I	II	III	IIII	卌
Medium	II	IIII	卌 I	卌 III	卌 卌
Large	III	卌 I	卌 IIII	卌 卌 II	卌 卌 卌

A 4, 8 **B** 9, 15 **C** 12, 30

15 brg, bgr, rbg, rgb, gbr, grb **16** **A** Blue **B** Blue **C** Blue

17

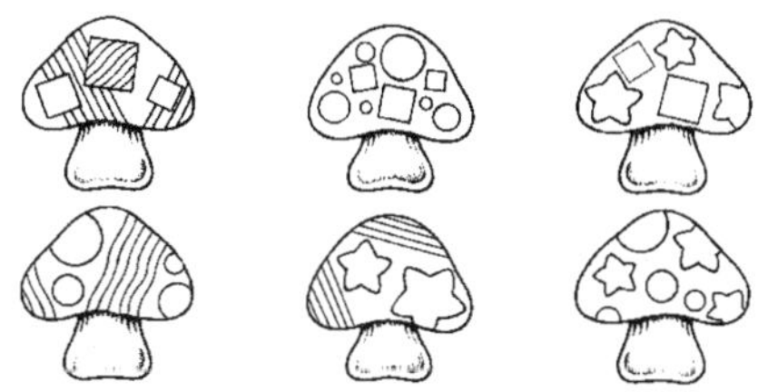

Challenge: 10 ways

18

Bushes	Bugs to eat in each bush	Bugs he can eat in all
A and B	2, 5	2 + 5 = 7
A and C	2, 3	2 + 3 = 5
A and D	2, 6	2 + 6 = 8
B and C	5, 3	5 + 3 = 8
B and D	5, 6	5 + 6 = 11
C and D	3, 6	3 + 6 = 9

19 puddle 3, rock 3, fern 2 **A** 5 **B** 6

20. lily pad 5, toadstool 2, grass 2 **A** 5 **B** 4

21

A 4 **B** 6 **C** 2 **D** 5

22

A 5 **B** 1 **C** 6

Look for a pattern

ACTIVITY 1

Name ____________________

Use small frogs.
Record your work.

The frogs are sitting on logs.
2 frogs sit on the first log.
4 frogs sit on the second log.
6 frogs sit on the third log.

A. If the frogs continue sitting in the same way, how many frogs will sit on the fourth log? _____

B. How many frogs in all? _____

Look for a pattern

ACTIVITY 2

Name ____________________

Use small frogs.
Record your work.

Fiona Frog is hopping across the river.
On her first hop, she lands on rock number 2.
On her second hop, she lands on rock number 4.

If she keeps hopping in the same way,
which rock will be her 6th hop?

Fill in the table. Look for a pattern.
It can help you solve the problem.

Hop	1st	2nd	3rd	4th	5th	6th
Rock	2	4	6			

Number of hops to cross the river. _____

DO MORE

Suppose there are 14 rocks across the river.
How many hops will Fiona take to cross the river?

Look for a pattern

Name ______________________

Activity 3

Use small frogs.
Record your work.

Freddie Frog is stuck in the well.

On day 1 he jumps up 1 step on the ladder.
On day 2 he jumps up 2 steps on the ladder.
On day 3 he jumps up 3 steps on the ladder.

Use the picture to show how Freddie jumps.

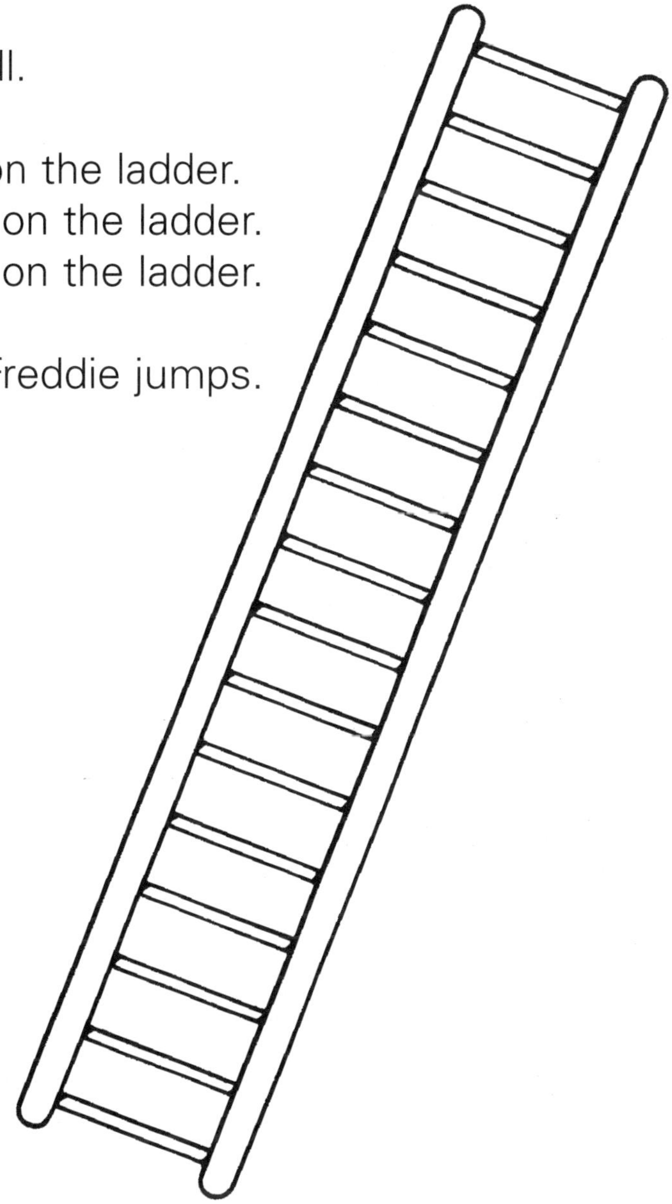

A. If he continues jumping in the same way, how many days will he need to reach the top? _____

B. Write a number sentence to show how many jumps he made in all. ______________________

Look for a pattern

ACTIVITY 4

Name_______________

Use small frogs.
Record your work.

The frogs are lining up in rows.
The first row has 1 frog.
The second row has 2 frogs.
The third row has 4 frogs.

If the frogs continue making rows in the same way, how many frogs will be in row 5?

Fill in the table. Look for a pattern.
It can help you solve the problem.

Row	Number of frogs
1	1
2	2
3	4
4	
5	

DO MORE

How many frogs in all?

Use a picture

Name________________________

ACTIVITY 5

Use small frogs.
Record your work.

Read the clues.
Find each frog's mailbox.

Blue Frog's box is above Green Frog's.
Yellow Frog's box is next to Blue Frog's.

Purple Frog's box is below Yellow Frog's.
Green Frog's box is next to Purple Frog's.
Orange Frog's box is in the top row.
Find Red Frog's box.

DO MORE

Find the frog with the most mail.
Find the frog with the least mail.
How many letters in all for Yellow and Red Frog?
How many more letters for Green Frog than for Purple Frog?

ACTIVITY 6

Name________________________

Use small frogs and the Lily Pad Number Line (page 31).
Record your work.

Place Green Frog on number 7.
Place Purple Frog on number 18.
Place Red Frog on number 12.

Use the clues.
Write a number sentence to show each frog's place.
Example: Red Frog is 5 hops ahead of Green Frog. 7 + 5 = 12

A. Blue Frog is 2 hops ahead of Purple Frog. ________________

B. Orange Frog is 2 hops behind Red Frog. ________________

C. Yellow Frog is 4 hops behind Green Frog. ________________

DO MORE

Use words to compare where 2 frogs are sitting.
Write the numbers sentence to compare the same frogs.

Use a picture

ACTIVITY 7

Name____________________

Use small frogs and the Lily Pad Number Line (page 31). Record your work.

A. Red Frog hops by threes. Write the numbers where it will land:

3, 6, 9, _____, _____, _____, _____, _____

B. Blue Frog hops by fours. Write the numbers where it will land:

4, 8, 12, _____, _____, _____, _____

C. Write the lily pad numbers where both frogs land.

_____ and _____

DO MORE

Yellow Frog hops by fives. Write the numbers on which it will land.
On which lily pad do both Yellow Frog and Red Frog hop?
On which pad do both Yellow Frog and Blue Frog hop?

Use a picture

ACTIVITY 8

Name________________________

Use small frogs and the Lily Pad Number Line (page 31).
Record your work.

Orange Frog hops by twos along the lily pad path.
She jumps to 2, then to 4, then to 6.

Green Frog hops by threes.
He jumps to 3 then to 6.

BOTH frogs land on lily pad 6!

If both frogs keep hopping the same way, on which lily pads will they both land?

_____, _____, and_____

ACTIVITY 9

Name_______________________

Use small frogs and the Lily Pad Number Line (page 31).
Record your work.

Place the frogs where they belong on your number line.

Green is on 7. Purple is on 18. Yellow is on 3.
Red is on 12. Blue is on 20. Orange is on 10.

Write a number sentence for each problem.

Example:
How many more hops does Yellow Frog need to meet Green Frog?
3 + 4 = 7 or 7 – 3 = 4

A. How many more hops does Red Frog need to meet Blue Frog?

B. How many hops back will Orange Frog need to take to sit with Green Frog?

C. How many more is Red Frog than Orange Frog?

D. How many more is Purple Frog than Green Frog?

Make a table

Name________________________

Use small, medium, and large frogs.
Record your work.

Use the clues.
Put each frog in the table where it belongs.

The frogs are red, yellow, and green.
There is one small frog, one medium frog, and one large frog in each color.

	Red	Yellow	Green
Small			
Medium			
Large			

A. How many frogs in all? _____

B. How many large frogs? _____

C. How many yellow frogs? _____

D. If 4 frogs hop away, how many are left? _____

Make a table

ACTIVITY 11

Name ______________________

Use small frogs.
Record your work.

Color the frogs.
Find how many frogs for each color.

Use tally marks to count each color.

Red	
Green	
Yellow	
Blue	

Write the number sentence to answer each question.

A. How many red frogs and blue frogs in all? ______________

B. How many green frogs and yellow frogs? ______________

C. How many more red frogs than blue frogs? ______________

D. How many frogs in all? ______________

Make a table

ACTIVITY 12

Name ____________________

Use small frogs.
Record your work.

Franny Frog is looking for a tasty bug.
She sees flies, beetles, and ants.
Help her count how many of each bug she sees.

Use tally marks in the table to show each kind of bug.

Flies	
Beetles	
Ants	

Write the number sentence to answer each question.

A. How many flies does she see? ____________________

B. How many beetles and ants does she see? ____________________

C. How many more beetles than flies does she see? ____________

Make a table

ACTIVITY 13

Name______________________

Use small frogs.
Record your work.

Freddie Frog wonders where to hop next.
Find all the different places he can hop.

He can hop on a

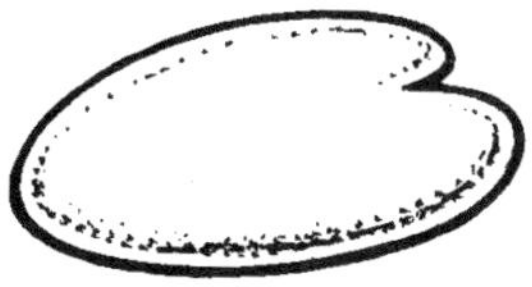 or  or 

In the –

 or 

Use the table.
Mark each place Ferdie can hop.

How many different places can he hop? _____

	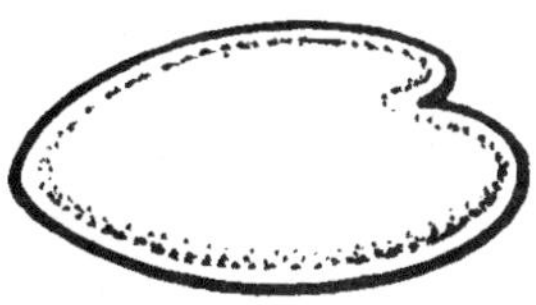		

Make a table

ACTIVITY 14

Name________________________

Use small, medium, and large frogs.
Record your work.

Small Frog croaks 1 time in one minute.
Medium Frog croaks 2 times in one minute.
Large Frog croaks 3 times in one minute.

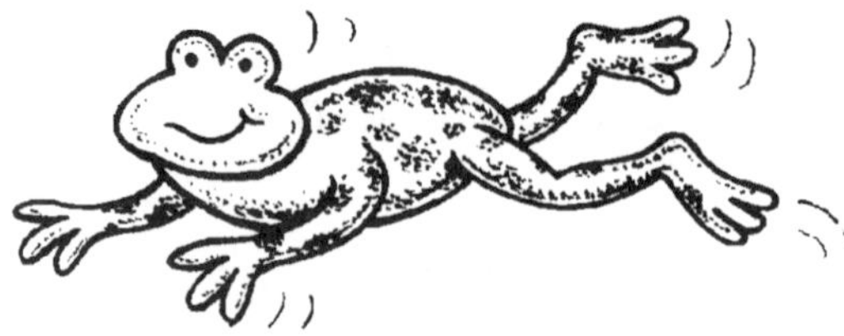

When the three frogs are together, how many croaks will you hear in one minute?

Use tally marks or numbers to fill in the table.
It can help you solve the problem.

	1 minute	2 minutes	3 minutes	4 minutes	5 minutes
Small	𝍷	𝍷𝍷	𝍷𝍷𝍷		
Medium	𝍷𝍷	𝍷𝍷𝍷𝍷			
Large	𝍷𝍷𝍷	𝍸 𝍷			

A. How many times will Medium Frog croak in 2 minutes? _____
In 4 minutes? _____

B. How many times will Large Frog croak in 3 minutes? _____
In 5 minutes? _____

C. How many times will ALL 3 frogs croak in 2 minutes? _____
In 5 minutes? _____

Make an organized list

ACTIVITY 15

Name ____________________

Use small frogs.
Record your work.

Blue Frog, Red Frog, and Green Frog are sitting on toadstools.
How many different ways can they sit?

Make a list to show the different ways.
Write b for blue, r for red, and g for green.

Here are 2 ways to get you started:

Way 1	b	r	g
Way 2	b	g	r
Way 3			
Way 4			
Way 5			
Way 6			

CHALLENGE!
Yellow Frog wants to join the group.
Draw a toadstool for Yellow Frog.
Find the different ways the four frogs can sit.

Make an organized list

ACTIVITY 16

Name ____________________

Use small frogs.
Record your work.

Red Frog, Purple Frog, Green Frog, and Blue Frog want to sit on the lily pad.
Only 2 frogs can sit on the lily pad at one time.

A. First, Red Frog and Purple Frog sit on the lily pad.
Then Red Frog and Green Frog.
Which other frog can sit with Red Frog?
Add that frog to the list:

Red, Purple
Red, Green
Red, ____________

B. Next, Purple Frog wants to sit with each frog.
She already sat with Red Frog.
She sits with Green Frog.
Which other frog can she sit next to?
Add that frog to the list:

Purple, Green
Purple, ____________

C. Green Frog already sat next to Red Frog and Purple Frog.
Which frog has Green Frog NOT sat next to?
Write your answer:

Green, ____________

Make an organized list

ACTIVITY 17

Name____________________

Record your work.

Fern is decorating toadstool tops.
She paints:

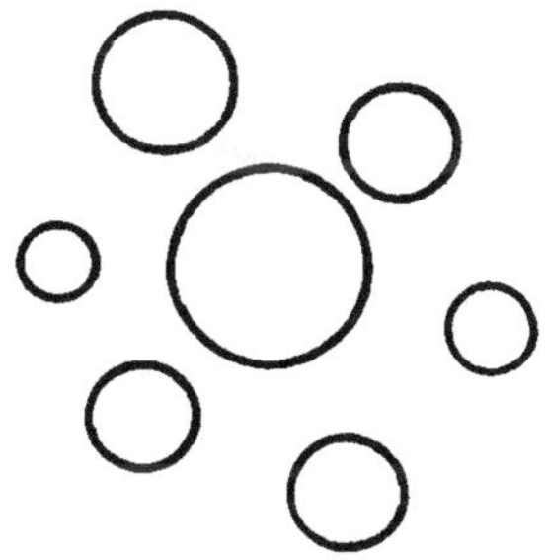

She puts two designs on each toadstool.
Help Fern finish her work.

Draw pictures to show all the ways.

CHALLENGE!
Add a new pattern.
Now how any different ways can Fern paint the toadstools?

Make an organized list

ACTIVITY 18

Name________________________

Use small frogs.
Record your work.

There are 4 bushes near Green Frog's home.
Green Frog jumps into 2 bushes.
He eats all the bugs in the 2 bushes.
How many bugs can he eat?

Use the table.
Write a number sentences to show all ways Green Frog can eat the bugs.
Example:
Green Frog jumps into bushes A and B: 2 + 5 = 7

Bushes	Bugs to eat in each bush	Bugs he can eat in all
A and B	2, 5	2 + 5 = 7
A and C	2, 3	2 + 3 = 5
A and D		
B and C		
B and D		
C and D		

Use logical thinking

ACTIVITY 19

Name___________________________

Use small frogs.
Record your work.

Use the clues.
Find how many frogs are in the puddle, on the rock, and under the fern.

There are 8 frogs in all.
3 frogs are in the puddle.
There are more frogs in the puddle than under the fern.
There is an odd number of frogs on the rock.

A. How many frogs are on the rock and under the fern? _____

B. How many frogs are in the puddle and on the rock? _____

Use logical thinking

ACTIVITY 20

Name______________________

Use small frogs.
Record your work.

Use the clues.
Find the number of frogs on the lily pad, on the toadstool, and in the grass.

There are 9 frogs.
Two frogs are in the grass.
There are more frogs on the lily pad than on the toadstool.
There is an even number of frogs on the toadstool.

A. How many frogs are on the lily pad? _____

B. How many frogs in all are on the toadstool and in the grass? _____

Use logical thinking

ACTIVITY 21

Name ____________________

Use small frogs.
Record your work.

Use the clues to color the frogs.

There are eight frogs.
Three frogs are purple.
There are no orange or blue frogs.
There is the same number of purple as yellow frogs.
There is the same number of green as red frogs.

Color the frogs.

A. How many yellow and green frogs? _____

B. How many purple and yellow frogs? _____

C. How many more purple than red frogs? _____

D. If the purple frogs hop away, how many will be left? _____

Use logical thinking

ACTIVITY 22

Name________________________

Use small frogs.
Record your work.

Use the clues to color the frogs.

There are 10 frogs.
There are two green frogs.
There is one more blue frog than green frogs.
There is the same number of green and red frogs.
There are no purple or orange frogs.

Color the frogs.

A. How many blue frogs and green frogs in all? _____

B. How many more red frogs than yellow frogs? _____

C. If the green frogs hop away, how many frogs are left? _____

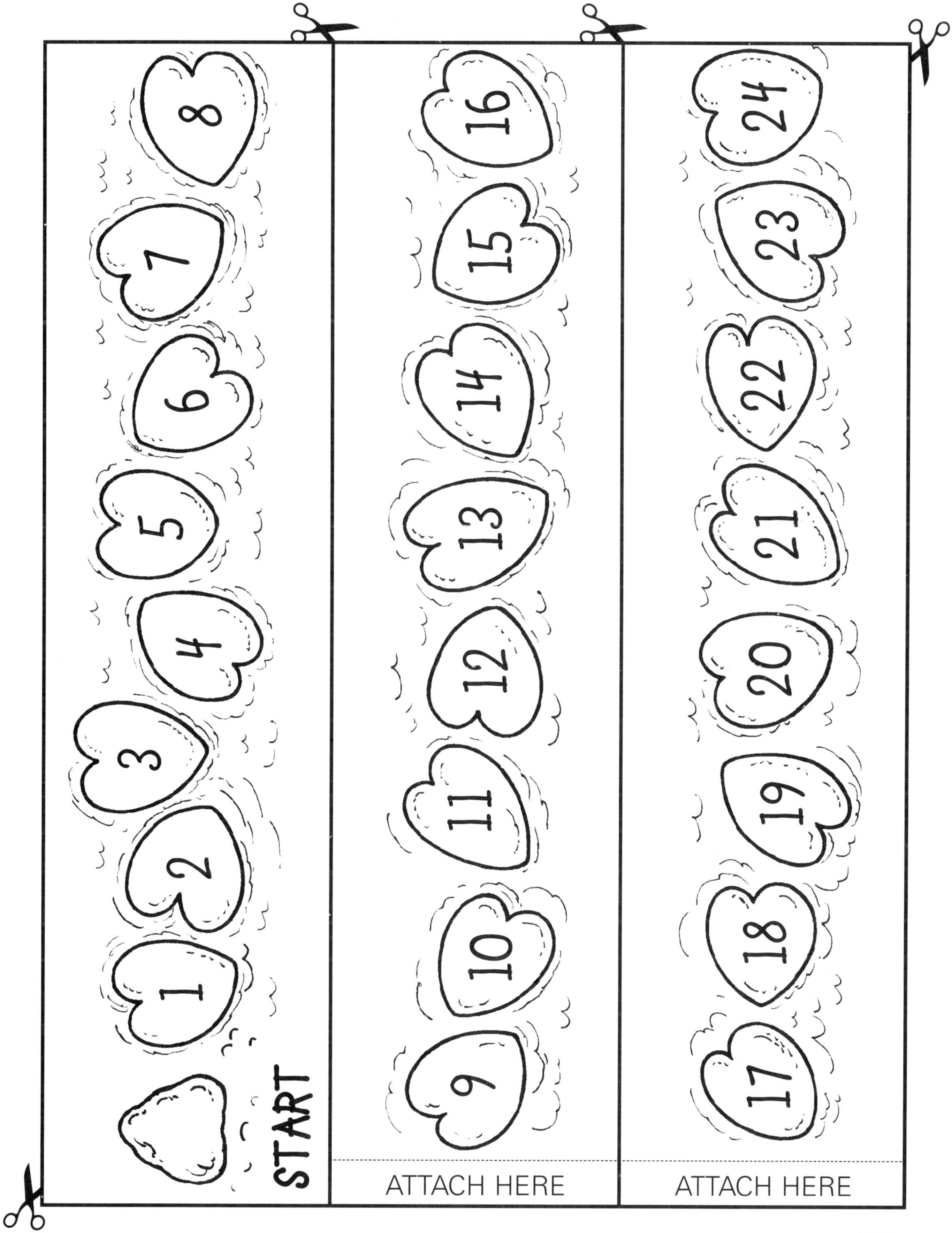
START
1
2
3
4
5
6
7
8
9
10
11
12
13
14
15
16
17
18
19
20
21
22
23
24
ATTACH HERE
ATTACH HERE

Name____________________

Write Your Own Problem

Write a Funtastic Frogs problem for your friends to solve!
Use frogs to help you.

____

____

____

____

____

____

························fold here to hide the answer!························

Show how YOU solved the problem:

The answer is: ________

